U.S. Army Anti-Guerrilla War

I0060039

U.S. Army Anti-Guerrilla Warfare Manual

ISBN: 978-1-942825-01-2

Author(s): U.S. Army

Editor: Kambiz Mostofizadeh

Publisher: Mikazuki Publishing House

Date Published: January 7th, 2015

Description: The U.S. Army Anti-Guerrilla Warfare Manual is an official manual used by the U.S. Army.

1

U.S. Army Anti-Guerrilla Warfare Manual

TABLE OF CONTENTS

U.S. Army Anti-Guerrilla Warfare Manual

CHAPTER 1

INTRODUCTION

Section I. GENERAL

Purpose and Scope

a. This manual provides guidance to the commanders and staffs of combined arms forces which have a primary mission of eliminating irregular forces.

b. The text discusses the nature of irregular forces comprised of organized guerrilla units and underground elements, and their supporters; and the organization, training, tactics, techniques, and procedures to be employed by a combined arms force, normally in conjunction with civil agencies, to destroy large, well-organized irregular forces in active or cold war situations.

c. These operations may be required in situations wherein an irregular force either constitutes the only enemy or threatens rear areas of regular military forces which are conducting conventional operations.

d. The material contained herein is applicable to both nuclear and non-nuclear warfare.

Terms

a. The term irregular, used in combinations such as irregular forces, irregular activities, and counter-irregular operations, is used in the broad sense to refer to all types of nonconventional forces and operations. It includes guerrilla, partisan, insurgent, subversive, resistance, terrorist, revolutionary, and similar personnel, organizations and methods.

b. Irregular activities include acts of a military, political, psychological, and economic nature, conducted predominantly by inhabitants of a nation for the purpose of eliminating or weakening the authority of the local government or an occupying power, and using primarily irregular' and informal groupings and measures.

Basic Precepts

a. An irregular force is the outward manifestation of a resistance movement against the local government by some portion of the population of an area. Therefore, the growth and continuation of an irregular force is dependent on support furnished by the population even though the irregular force also receives support from an external power.

b. When an irregular force is in its formative stage it may be eliminated by the employment of civil law enforcement measures and removal of the factors which cause the resistance movement. Once formed, an irregular force is usually too strong to be eliminated by such measures. A stronger force, such as a military unit, can destroy the irregular force, but the resistance movement will, when convinced that it is militarily feasible to do so, reconstitute the irregular force unless the original causative factors are also removed or alleviated.

c. The ultimate objective of operations against an irregular force is to eliminate the irregular force and prevent its resurgence. To attain this objective the following tasks must be accomplished:

(1) The establishment of an effective intelligence system to furnish detailed, accurate, and current knowledge of the irregular force.

(2) The physical separation of guerrilla elements from each other, their support base in the local population, underground elements, and any sponsoring power.

(3) The destruction of irregular force elements by the defection,

surrender, capture, or death of individual members.

(4) The provision of political, economic, and social necessities and the ideological reeducation of dissident elements of the population to present resurgence of the irregular force.

Principles of Operation

(1) Direction of the military and civil effort on this level is vested in a single authority, either military or civil.

(2) Military operations are conducted in conjunction with specified civil rights, liberties, and objectives.

(3) Operations are assumed to be predominantly offensive operations.

(4) Police, Civil operations, and civic action operations are conducted simultaneously.

5) Anti-Guerrilla forces employed against guerrilla elements are organized to have a higher degree of aggressiveness and mobility than the guerrilla elements.

OPERATIONAL ENVIRONMENT

Ideological Basis for Resistance

a. The fundamental cause of large-scale resistance movements stems from the dissatisfaction of some portion of the population, whether real, imagined, or incited, with the prevailing political, social, or economic conditions. This dissatisfaction is usually centered around a desire for one or more of the following:

(1) National independence.

(2) Relief from actual or alleged oppression.

(3) Elimination of foreign occupation or exploitation.

(4) Economic and social improvement.

(5) Elimination of corruption.

U.S. Army Anti-Guerrilla Warfare Manual

Religious Expression.

a). "In-country" factors may cause a resistance movement to form locally or a resistance movement may be inspired by "out-of -country" elements who create and sponsor such a movement as a means of promoting their own cause. Often, another country will lend support to a local resistance movement and attempt to control it to further its own aims.

b. Resistance movements begin to form when dissatisfaction occurs among strongly motivated individuals who cannot further their cause by peaceful and legal means. Under appropriate conditions, the attitudes and beliefs of these individuals, who are willing to risk their lives for their beliefs, spread to family groups and neighbors. The population of entire areas may soon evidence widespread discontent. When legal recourse is denied, discontent grows into disaffection and members of the population participate in irregular activities.

U.S. Army Anti-Guerrilla Warfare Manual

c. Small dissident groups living and working within the established order gradually organize into underground elements that conduct covert irregular activities. As members of underground organizations are identified and as the spirit of resistance grows, overt guerrilla bands form in secure areas and become the military arm of the irregular force. Characteristically, the scope of irregular activities progress in this order: Passive objection, individual expression of opposition, minor sabotage, major sabotage, individual violent action, and organized group violent action.

d. Once individuals have participated in irregular activities, should there be any change of heart, they are usually forced to continue, either by pressure from their comrades or by being designated criminals by local authority.

U.S. Army Anti-Guerrilla Warfare Manual

Irregular Force Organization

a. The organization of irregular forces varies according to purpose, terrain, character and density of population, availability of food, medical supplies, arms and equipment, quality of leadership, amount and nature of external support and direction, and the countermeasures used against them. Units or elements may vary in size from a few saboteurs to organized paramilitary units of division size or larger with extensive support organizations.

b. A large irregular force normally consists of two organized elements: a guerrilla element which operates overtly, and an underground element which operates covertly. Members of large guerrilla units are usually severed from their normal civilian pursuits while members of small guerrilla bands may alternately be either guerrillas or apparently peaceful citizens. Members of the underground usually maintain their civilian pursuits. Elements are usually supported by individuals and small groups who may or may not be formal members of either element but who furnish aid in intelligence, evasion and escape, and supplies. Such supporters are often considered to be a part of the underground. A given individual may be a member of two or more organizations and may participate in many forms of irregular activity.

c. The underground elements of an irregular force must conduct the majority of their activities in a covert manner because of the countermeasures used against them. They are usually found in all resistance areas. Successful organizations are compartmented by cells for security reasons. The cellular organization prevents one member, upon capture, from compromising the entire organization.

d. Organized guerrilla units are usually found in areas where the

U.S. Army Anti-Guerrilla Warfare Manual

terrain minimizes the mobility, surveillance, and firepower advantage of the opposing force. They vary from small groups who are lightly armed, to large paramilitary units of division size or larger with extensive support organizations. Large organizations normally include elements for combat, assassination and terrorism, intelligence and counterintelligence, civilian control, and supply. Guerrilla units are composed of various categories of personnel. Members may include:

(1) Civilian volunteers and those impressed by coercion.

(2) Military leaders and specialists.

(3) Deserters.

(4) In time of active war, military individuals or small groups such as those who have been cut off, deliberate stay-behinds, escaped prisoners of war, and downed airmen.

U.S. Army Anti-Guerrilla Warfare Manual

Irregular Force Activities and Tactics

a. An irregular force presents an elusive target, since it will usually disperse before superior opposition, and then reform to strike again. However, as the guerrilla elements of an irregular force grow and approach parity with regular units in organization, equipment, training, and leadership, their capabilities and tactics likewise change and become similar to those of a regular unit.

(1) Overt irregular activities include acts of destruction against public and private property, transportation and communications systems ; raids and ambushes against military and police headquarters, garrisons, convoys, patrols, and depots; terrorism by assassination, bombing, armed robbery, torture, mutilation, and kidnaping; provocation of incidents, reprisals, and holding of hostages ; and denial activities, such as arson, flooding, demolition, use of chemical or biological agents, or other acts designed to prevent use of an installation, area, product, or facility.

U.S. Army Anti-Guerrilla Warfare Manual

(2) Covert irregular activities include — espionage, sabotage, dissemination of propaganda and rumor;;, delaying or misdirecting orders, issuing false or misleading orders or reports, assassination, extortion, blackmail, theft, counterfeiting, and identifying individuals for terroristic attack.

Type Guerrilla Organization

Irregular force tactics vary; however, the following tactics are common to all irregular forces:

(1) Guerrilla tactics. The tactics used by the guerrilla are designed to weaken his enemy and to gain support of the population. Guerrilla tactics follow well-known precepts: If the enemy attacks, "disappear;" if he defends, "harass;" and if he withdraws or at any time he is vulnerable, "attack."

(a) Guerrilla tactics are primarily small unit, infantry-type

U.S. Army Anti-Guerrilla Warfare Manual

tactics which make full use of accurate intelligence, detailed planning and rehearsal, simple techniques of maneuver, speed, surprise, infiltration, specialization in night operations, and the undermining of enemy morale. Speed is a relative thing and is usually accomplished by such actions as marching rapidly 2 or 3 nights to reach the area of attack. Surprise is gained by the combined elements of speed, secrecy, selection of unsuspected objectives, and deliberate deception. Infiltration is a basic tactic of successful guerrilla units and they quickly develop great skill in infiltrating areas occupied by military units. By specializing in night operations, a guerrilla force effectively reduces its vulnerability to air and artillery attack. Enemy morale is undermined by constant harassment, exhibition of a violent combative spirit, fanaticism, self-sacrifice,
and extensive use of propaganda, threats, blackmail, and bribery.

U.S. Army Anti-Guerrilla Warfare Manual

(2) Artillery and armor are seldom available to guerrilla units. This frees the guerrilla force from extensive combat trains, permits excellent ground mobility in rugged terrain, and facilitates infiltration techniques. When artillery is available, individual pieces are usually emplaced at night, dug-in, and expertly camouflaged.

(a) When surrounded or cut off, guerrillas immediately attempt to break out by force at a single point or disband and exfiltrate individually. If both fail, individual guerrillas attempt to hide or mingle with the peaceful civilian population.

(b) Guerrilla base areas are carefully guarded by a warning net consisting of guerrillas and/or civilian supporters. Warning stations, which are relocated frequently, cover all avenues of approach into the guerrilla area.

(3) Underground element tactics. The tactics employed by underground elements are designed to gain the same results as guerrilla tactics.

(a) Underground organizations attempt, through nonviolent persuasion, to indoctrinate and gain the participation of groups of the population who are easily deceived by promises and, through coercion by terror tactics, to force others to participate.

(b) Espionage and sabotage are common to all underground groups.

(c) Terroristic attack such as assassination and bombings are used to establish tension and reduce police or military control.

(d) Propaganda is disseminated by implanting rumors, distributing leaflets and placards, and when possible, by operating clandestine radio stations for broadcasting purposes.

(e) Agitation tactics include fostering of black markets, and promoting demonstrations, riots, strikes, and work slow-down.

(f) Overt and covert activities. Both overt and covert activities are employed in active war or occupation situations to intentionally draw off enemy combat troops from commitment to their primary mission.

Political Factors.

a. Operations against irregular forces are sensitive politically, particularly during cold war operations. The scope and nature of missions assigned and commanders' decisions will emphasize political aspects to a greater extent than in normal military operations.

b. The presence of a foreign military force operating against irregular forces will invariably be exploited by hostile political propaganda.

19

U.S. Army Anti-Guerrilla Warfare Manual

Geographical Factors

a. The vastness of an area over which such operations may be conducted can result in large areas which cannot be closely controlled.

When such areas are sparsely populated or when they contain unfriendly elements of the population, they become likely areas for the development of a hostile resistance movement.

1), Areas of rugged or inaccessible terrain, such as mountains, forests, jangles, and swamps, are extremely difficult to control, and the guerrilla elements of an irregular force are most likely to flourish in such areas.

Relationship of Forces

a. Under the Geneva Conventions, as discussed in FM 27-10, "The Law of Land Warfare," a guerrilla may, in time of war, have legal status ; when he is captured, he is entitled to the same treatment as a regular prisoner of war if he:

U.S. Army Anti-Guerrilla Warfare Manual

(1) Is commanded by a person responsible for his subordinates.

(2) Wears a fixed and distinctive sign recognizable at a distance.

(3) Carries arms openly.

(4) Conducts operations in accordance with the laws and customs of war.

(5) The underground elements of an irregular force normally do not hold legal status.

(6) Groups of civilians that take up arms to resist an invader have the status of belligerents, but inhabitants of occupied territory who rise against the occupier are not entitled to belligerent status. However, the occupier has the responsibility of making his occupation actual and effective by preventing organized resistance, and

promptly suppressing irregular activities. The law of land warfare further authorizes the occupier to demand and force compliance with counter-measures (FM 27-10).

a. Historically, legal status has been of little concern to members of an irregular force and has had little effect on their decision to participate in irregular activities.

CHAPTER 2

CONCEPT AND CONDUCT OF OPERATIONS

Section I. COMMAND AND CONTROL

Active War

a. The measures which U.S. military commanders may take against irregular forces during hostilities and in occupied enemy territory are limited to those which are authorized by the laws of land warfare

b. Army operations against irregular forces in a general or limited war will be conducted within the command structure established for the particular theater. The senior headquarters conducting the operation may be joint, combined, or uni-service. If an established communication zone exists, control of operations against an irregular force in this area will normally be a responsibility of the Theater Army Logistical Command or the appropriate section headquarters.

c. Continuity of commanders and staff officers and retention of the same units within an area are desirable to permit commanders, staffs, and troops to become thoroughly acquainted with the terrain, the local population, the irregular force organization, and its techniques of operations.

U.S. Army Anti-Guerrilla Warfare Manual

Cold War Situations

a. In liberated areas in which a friendly foreign government has been reestablished and in sovereign foreign countries in time of peace, the authority which U.S. military commanders may exercise against irregular forces is limited to that permitted by the provisions of agreement that are concluded with responsible authorities of the sovereign government concerned.

b. The chief of the U.S. diplomatic mission in each country, as the representative of the President, is normally the channel of authority on foreign policy and the senior coordinator for the activities of all representatives of U.S. agencies and services in such a country. In some instances, diplomatic representation may not be present, or the relationship between the diplomatic representative and the military commander may be further delineated by executive order.

U.S. Army Anti-Guerrilla Warfare Manual

Coordination 1

Equipment and Advisory support

Possible relationships in a cold war situation.

c. Uniservice headquarters, or joint or combined commands may control operations against irregular forces in a cold war situation. Participation by the host country is normal and usually makes a combined command mandatory.

a. Responsibility for civil control and administration is specified in agreements reached with the host country and usually is vested in the legally constituted government to the maximum extent possible. If the military force commander has been given civil responsibilities, agreements usually will state that full responsibility for liberated or secured areas be transferred to local authorities as soon as the military situation permits.

U.S. Army Anti-Guerrilla Warfare Manual

Section II. PLANNING

General

a. Operations against irregular forces are designed to establish control within the resistance area, eliminate the irregular force, and assist in the reconstruction, habilitation, and reeducation required to provide a suitable atmosphere for peaceful living. These goals are sought concurrently, but in specific instances priorities may be established. The specific actions required to attain each goal are often the same, and even when different, are usually planned and conducted concurrently.

Operations against irregular forces are planned according to these basic considerations.

(1) The majority of operations consist of small unit actions.

(2) Operations are primarily offensive in nature; once initiated, they are continued without halt to prevent

U.S. Army Anti-Guerrilla Warfare Manual

irregular force reorganization and resupply. Lulls in irregular activities or failure to establish contact with hostile elements may reflect inadequate measures in the conduct of operations rather than complete success.

(3) Operations are designed to minimize the irregular force's strengths and to exploit their weaknesses.

{a) The greatest strength of an irregular force lies in its inner political structure and identification with a popular cause, its ability to conceal itself within the civil population, the strong motivation of its members, and their knowledge of the resistance area.

(b) The greatest weakness of an irregular force lies in its dependence upon support by the civil population; its lack of a reliable supply system for food, arms, and ammunition; aid its lack of transportation, trained leadership, and communications.

(4) The close relationship between the civil population and the irregular force may demand enforcement of stringent control measures. In some cases it may be necessary to relocate entire villages, or to move individuals from outlying areas into population centers. It may be necessary to relocate those who cannot be protected from guerrilla attack, and those who are hostile and can evade control.

a. Terrain and the dispositions and tactics of guerrilla forces usually limit the effectiveness of artillery. However, the demoralizing effect of artillery fire on guerrillas often justifies its use even though there is little possibility of inflicting material damage. Ingenuity and a departure from conventional concepts often make artillery support possible under the most adverse circumstances.

b. The rough terrain normally occupied by guerrilla forces often limits the use of armored vehicles. However, armored vehicles provide protected communications, effective mobile roadblocks, and convoy escort. Planning should therefore include the employment of armor whenever its use is feasible. When used against guerrillas, armored vehicles must be closely supported by infantry, as guerrillas are skilled at improvising antitank means and may be equipped with recoilless weapons and light rockets.

c. The helicopter has wide application when used against irregular forces, subject to the usual limitations of weather and visibility. Its capability for delivering troops, supplies, and suppressive fires, and its ability to fly at low speed, to land in a small clearing, and to hover make it highly useful.

d. Morale of forces engaged in operations against irregular forces presents some planning considerations not encountered in other types of combat. Operations against a force that seldom offers a target, disintegrates before opposition, and then re-forms and strikes again where it is least expected may induce strong feelings of futility among soldiers and dilute their sense of purpose.

e. Activities between adjacent commands must be coordinated to insure unity of effort. If a command in one area carries out vigorous operations while a neighboring command is passive, guerrilla elements will move into the quiet area until danger has passed. Underground elements will either remain quiet or transfer their efforts temporarily.

f. Definite responsibilities for the entire area of operations are specified, using clearly defined boundaries to subdivide the area. Boundaries should not prevent hot pursuit of irregular force elements into an adjacent area. Coordination should, however, be accomplished between affected commands at the earliest practicable opportunity. Boundaries should be well defined and should not intersect swamps, dense forests, mountain ridges, or other key terrain features used by guerrilla elements for camps, headquarters, or bases. Similarly, well-defined boundaries should be used to divide urban areas to insure complete coverage.

g. Terrain and the dispositions and tactics of guerrilla forces furnish excellent opportunity for the employment of chemical and biological agents and riot control agents. Operations against irregular forces should evaluate the feasibility of chemical and biological operations to assist in mission accomplishment.

Planning Factors

a. Planning for operations against irregular forces requires a detailed analysis of the area concerned and its population. Close attention is given to both the civil (political, economic, social) and the military situations.

U.S. Army Anti-Guerrilla Warfare Manual

b. The following specific factors are considered in the commander's estimate:

(1) The motivation and loyalties of various segments of the population, identification of hostile and friendly elements, vulnerability of friendly or potentially friendly elements to coercion by terror tactics, and susceptibility to enemy and friendly propaganda. Particular attention is given to the following:

(a) Farmers and other rural dwellers.

(b) Criminal and tough elements.

(c) Adherents to the political philosophy of the irregular force or to similar philosophies.

(d) Former members of armed forces.

(e) Existence of strong personalities capable of organizing an irregular force and their activities.

(2) The existing policies and directives regarding legal status and treatment of civil population and irregular force members.

(3) The terrain and weather to include

(a) The suitability of terrain and road and trail net for both irregular force and friendly force operations.

(b) The location of all possible hideout areas for guerrillas.

(c) The location of possible drop zones and fields suitable for the operation of aircraft used in support of guerrilla units.

(4) The resources available to the irregular force, including:

{a) The capability of the area to furnish food.

(b) The capability of friendly forces to control the harvest, storage, and distribution of food.

(c) The availability of water and fuels.

(d) The availability of arms, ammunition, demolition materials, and other supplies.

(5) Irregular force relations with any external sponsoring power, including:

(a) Direction and coordination of irregular activities.

(b) Communication with the irregular force,

(c) Capability to deliver organizers and supplies by air, sea, and land.

(6) The extent of irregular force activities and the force organization to include:

{a) Their origin and development.

(b) Their strength and morale,

(c) The personality of the leaders.

(d) Relations with the civil population.

(e) Effectiveness of organization and unity of command.

(f) Status of equipment and supplies.

(g) Status of training.

Effectiveness of communications.

Effectiveness of intelligence, including counterintelligence.

(a) The size and composition of forces available for counter-operations to include:

(a) Own forces.

(b) Other military units within the area.

(c) Civil police, paramilitary units, and self-defense units.

(8) The communication facilities available to allow effective control of forces engaged in counter operations.

Section III. PROPAGANDA AND CIVIC ACTION

Propaganda

a. Propaganda is planned and employed in the campaign to achieve the following immediate goals:

(1) Divide, disorganize, and induce defection of irregular force members.

(2) Reduce or eliminate civilian support of guerrilla elements.

U.S. Army Anti-Guerrilla Warfare Manual

(3) Dissuade civilians from participating in covert activities on the side of the irregular force.

(4) Win the active support of non-committed civilians.

(5) Preserve and strengthen the support of friendly civilians.

(6) Win popular approval of the local presence of friendly military forces.

(7) Obtain national unity or disunity as desired.

a. Propaganda activities aimed at achieving the immediate goals cited above must, as a minimum, be in consonance with each of the desired long-range goals, and should where possible contribute to their attainment.

U.S. Army Anti-Guerrilla Warfare Manual

a. For purposes of planning and directing the propaganda program, the population is divided into five target audiences. These are:

(1) Guerrilla units.

(2) Underground elements.

(3) Those civilians who provide information, supplies, refuge, and other assistance to the guerrillas and the underground.

(4) The non-committed civil population.

(5) Friendly elements of the civil population.

a. Propaganda themes are based on the recognizable aspects of friendly economic and political programs and on potentially divisive characteristics of hostile target audiences. Possible divisive factors are:

U.S. Army Anti-Guerrilla Warfare Manual

(1) Political, social, economic, and ideological differences among elements of the irregular force.

(2) Rivalries between irregular force leaders.

(3) Danger of betrayal.

(4) Extravagant living conditions of guerrilla elements.

(5) Scarcity of arms and supplies.

(6) Selfish motivation of opportunists and apparent supporters of the resistance movement.

a. Troop units of the TOE 3 should be included in the friendly force structure on a selective basis.

U.S. Army Anti-Guerrilla Warfare Manual

Civic Action

Civic action is any action performed by the military force utilizing military manpower and material resources in cooperation with civil authorities, agencies, or groups is designed to secure the economic or social betterment of the civilian community. Civic action can be a major contributing factor to the development of favorable public opinion and in accomplishing the defeat of the irregular force. Military commanders are encouraged to participate in local civic action projects wherever such participation does not seriously detract from accomplishment of their primary mission.

1) Civic action can include assistance to the local population as construction or rehabilitation of transportation and communication means, schools, hospitals, and churches; assisting in agricultural improvement programs, crop planting, harvesting, or processing; and furnishing emergency food, clothing, and medical aid as in periods of natural disaster.

U.S. Army Anti-Guerrilla Warfare Manual

a. Civic action programs are often designed to employ the maximum number of civilians until suitable economy is established. The energies of civilians' should be directed into constructive channels and toward ends which support the purpose of the campaign. Unemployed and discontented masses of people, lacking the bare necessities of life, are a constant hindrance and may preclude successful accomplishment of the mission.

b. Civic action is an instrument for fostering active civilian opposition to the irregular force and active participation in and support of opposition. The processes for achieving an awareness in civilian populations of their obligation to support stated aims should begin early. Commanders should clearly indicate that civil assistance by the military unit is not simply a gift but is also action calculated to enhance the civilians' ability to support the government.

c. Civil affairs units are employed to assist in the conduct of civic action projects and in the discharge of civil responsibilities.

Section IV. POLICE OPERATIONS

General

(Commanders and troops will usually be required to deal with local civil authorities and indigenous military personnel. When the local civil government is ineffective, the military commander may play a major role in keeping- order.

a. The activities and movements of the civil population are restricted and controlled, as necessary, to maintain law and order and to prevent a guerrilla force from mingling with and receiving support from the civil population. When the military commander is not authorized to exercise direct control of civilians, he must take every legitimate action to influence the appropriate authorities to institute necessary measures. Police-type operations may be conducted by either military or civil forces or a combination of both.

b. Restrictive measures are limited to those which are absolutely essential and can be enforced. Every effort is made to gain the willing cooperation of the local population to comply with controls and restrictions. However, established measures are enforced firmly and justly. Free movement of civilians is normally confined to their local communities. Exceptions should be made to permit securing food, attendance at public worship, and necessary travel in the event of illness.

c. Punishment of civilians, when authorized, must be used with realistic discretion. If the people become frustrated and alienated as a result of unjust punishment, the purpose is defeated. Care is taken to punish the true offender, since subversive acts are often committed to provoke unjust retaliation against individuals or communities cooperating with friendly forces.

U.S. Army Anti-Guerrilla Warfare Manual

d. Police operations employ roadblocks and patrol, search and seizure, surveillance and apprehension techniques. The following list is representative of the police-type controls and restrictions which may be necessary.

(1) Prevention of illegal political meetings or rallies.

(2) Registration and documentation of all civilians.

(3) Inspection of individual identification documents, permits, and passes.

(4) Restrictions on public and private transportation and communication means.

(5) Curfew.

(6) Censorship.

(7) (Controls on the production, storage, and distribution of foodstuffs and protection of food-producing areas.

(8) Controlled possession of arms, ammunition, demolitions, drugs, medicines, and money.

a. Patrolling is conducted to maintain surveillance of urban and rural areas, both night and day. Foot patrols are similar to normal police patrols, but are usually larger. Motor patrols are used to conserve troops and to afford speed in reacting to incidents. Aircraft are employed to maintain overall surveillance and to move patrols rapidly. Improvised landing pads such as roofs of buildings, parks, vacant lots, and streets are used. A patrol is a show of force and must always reflect high standards of precision and disciplined bearing. Its actions are rigidly controlled to preclude unfavorable incidents.

U.S. Army Anti-Guerrilla Warfare Manual

b. Surveillance of persons and places is accomplished, both night and day, by using a combination of the techniques employed by military and civil police and counterintelligence agencies. The majority of surveillance activities are clandestine in nature and may require more trained agent personnel than are normally assigned or attached to army units. Surveillance operations must be highly coordinated among all participating agencies.

c. Facilities for screening civilian and interrogating suspects are established and maintained. The requirements are similar to those for handling refugees, line crossers, and prisoners of war. Organization is on a team basis and normally consists of military and civil police, interpreter/translators, POW interrogators, and CI agents. Teams capable of operating at temporary locations on an area basis to support operations against civil disturbances and large scale search and seizure operations.

U.S. Army Anti-Guerrilla Warfare Manual

Roadblocks

Roadblocks are employed to control illegal possession and movement of goods, to check the adequacy of other controls, and to apprehend members of the irregular force.

a. Roadblocks are designed to halt traffic for search or to pass traffic as desired. They are established as surprise checks because irregular force members and their sympathizers soon devise ways of circumventing permanent checkpoints.

b. Teams are trained and rehearsed so as to be capable of establishing roadblocks in a matter of minutes at any hour. Local officials may be called on for assistance, to include the use of official translators, interrogators, and local women to search women and girls.

c. Roadblocks are established at locations which have suitable areas for assembling people under guard and for parking vehicles for search. Troops are concealed at the block and along the paths and roads leading to the block for the purpose of apprehending those attempting to avoid the block.

d. The inspection of documents and the search of persons and vehicles must be rapid and thorough. The attitude of personnel performing these duties must be impersonal and correct because many of the people searched will be friendly or neutral.

e. The establishment of roadblocks must be coordinated, but knowledge of where and when must be closely controlled.

Search and Seizure Operations

Search and seizure operations are conducted to screen a built-up area, apprehend irregular force members, and uncover and seize illegal arms, communication means, medicines, and supplies. Search and seizure operations may be conducted at any hour of night or day.

a. A search and seizure operation is intended to be a controlled inconvenience to the population concerned. The persons whose property is searched and whose goods are seized should be irritated and frightened to such an extent that they will neither harbor irregular force members nor support them in the future. Conversely, the action must not be so harsh as to drive them to collaboration with the irregular force because of resentment.

b. A built-up area to be searched is divided into block square zones. A search party is assigned to each zone and each party is divided into three groups: a search group to conduct the search, a security group to encircle the area to prevent ingress/egress, apprehend and detain persons, and secure the open street areas to prevent all unauthorized movement ; and a reserve group to assist as needed. The population may or may not be warned to clear the streets and to remain indoors until permission is given to leave. Force is used as needed to insure compliance. Each head of household or business is directed to assemble all persons in one room and then to accompany the search party to forestall charges of looting or vandalism.

c. Buildings which have been searched are clearly marked by a coded system to prevent false clearances.

Block Control

Block control is the constant surveillance and reporting of personnel movements within a block or other small populated area by a resident of that block or area who has been appointed and is supervised by an appropriate authority. Because of the civil authority and lengthy time required to establish a block control system, it is normally instituted and controlled by civil agencies. An

U.S. Army Anti-Guerrilla Warfare Manual

established block control system should be supported by the military force, and in turn, be exploited for maximum benefit.

a. Block control is established by dividing each block or like area into zones, each of which includes all the buildings on one side of a street within a block. A resident zone leader is appointed for each zone, and a separate resident block leader is appointed for each block. Heads of households and businesses in each zone are required to report all movements of people to the zone leader; to include arrivals and departures of their own families or employees, neighbors, and strangers. Zone leaders report all movements in their zone to the block leader. The block leader reports daily, to an appointed authority, on normal movements; the presence of strangers and other unusual circumstances are reported immediately.

b. The cooperation of leaders is secured by appealing to patriotic motives, by pay, or through coercion.

c. Informants are established separately within each block to submit reports as a check against the appointed block and zone leaders. An informant net system is established using key informants for the covert control of a number of block informants.

Installation and Community Security

Critical military and civil installations and key communities must be secured against sabotage and guerrilla attack.

a. Special attention is given to the security of food supplies, arms, ammunition, and other equipment of value to the irregular force. Maximum use is made of natural and manmade obstacles, alarms, illumination, electronic surveillance devices, and restricted areas. Fields of fire are cleared and field fortifications are constructed for guards and other local security forces. The local security system is supplemented by vigorous patrolling using varying schedules and routes. Patrolling distance from the installation or community is at least that of light mortar range. Specially trained dogs may be used with guards and patrols. As a defense against espionage and sabotage within installations, rigid security measures are enforced on native labor, to include screening, identification, and supervision. All security measures are maintained on a combat basis, and all personnel keep their weapons available for instant use. The routine means of securing an installation are altered frequently to prevent irregular forces from obtaining detailed accurate information about the composition and habits of the defense.

b. The size, organization, and equipment of local security forces are dictated by their mission, the size, composition, and effectiveness of the hostile force, and the attitude of the civil population. Security detachments in remote areas normally are larger than those close to supporting forces. Patrol requirements also influence the size of security detachments. Remote detachments maintain a supply level to meet the contingency of isolation from their base for long periods and should be independent of the local population for supply. Balanced against the desirability for self-sufficient detachments is the certainty that well-stocked remote outposts will be considered as prime targets for guerrilla forces as a source of arms and ammunition.

Reliable communications in the headquarters and security detachments are essential.

 a. Outlying installations are organized and prepared for all-round defense. Adequate guards and patrols are used to prevent surprise.

 b. Precautions are taken to prevent guards from being surprised and overpowered before they can give the alarm. Concealed and covered approaches to the position are mined and booby trapped, with due consideration for safety of the civil population. Areas from which short-range fire can be placed on the position are cleared and mined.

c. Personnel are provided with auxiliary exits and covered routes from their shelters to combat positions. Buildings used for shelters are selected with care. Generally, wooden or other light structures are avoided. If they must be used, the walls are reinforced for protection against small-arms fire. Supplies are dispersed and placed in protected storage. Adequate protection is provided for communication installations and equipment Individual alertness is maintained by frequent practice alerts which may include full scale rehearsal of defense, plans. Civilians, including children, are not permitted to enter the defensive positions. Civilian informants and observation and listening posts are established along routes of approach to the installations.

Security of Surface Lines of Communication

Surface lines of communication which have proved particularly vulnerable to guerrilla attack and sabotage should be abandoned, at least temporarily, if at all possible. Long surface lines of communication cannot be completely protected against a determined irregular force without committing an excessive number of troops. When a railroad, canal, pipeline, or highway must be utilized, the following measures may be required.

a. Regular defensive measures are taken to protect control and maintenance installations, repair and maintenance crews, and traffic. Vulnerable features such as major cuts in mountain passes, under-passes, tunnels, bridges, locks, pumping stations, water towers, power stations, and roundhouses require permanent guards or continuous surveillance of approaches. If necessary, the right-of-way of high-ways, railroads, canals, and pipelines are declared and posted as a restricted zone. The zone includes the area 300 meters on each side of the right-of-way. Civilian inhabitants are evacuated from the zone; underbrush is cleared and wooded areas are thinned to permit good visibility.

b. Frequent air and ground patrols are made at varying times, night and day, along the right-of-way, to discourage trespassing and to detect mines, sabotage, and hostile movements.

Armored vehicles, aircraft, and armored railroad cars are used by patrols when appropriate.

c. Lone vehicles, trains, and convoys which cannot provide their own security are grouped if practicable and are escorted through danger areas by armed security detachments. All traffic is controlled and reported from station to station.

Apprehended Irregular Force Members

a. Operations against irregular forces may generate sizable groups of prisoners which can create serious problems for both the military force and civil administration. Large numbers of confined personnel can generate further political, social, and economic difficulties. Therefore, the evaluation and disposition of prisoners must contribute to the ultimate objective.

b. When irregular force members defect or surrender voluntarily, they have, indicated that their attitudes and beliefs have changed, at least in part, and that they will no longer participate in irregular activities.

(1) Confinement should be temporary, only for screening and processing, and be separate from prisoners who have not exhibited a change in attitude.

U.S. Army Anti-Guerrilla Warfare Manual

(2) Promises made to induce defection or surrender must be met.

(3) Post-release supervision is essential but need not be stringent.

(4) Relocation may be required to prevent reprisal from former comrades.

a. When irregular force members are captured, they can be expected to retain the attitude which prompted their participation in irregular activities.

(5) Confinement is required and may continue for an extended period.

(6) Prisoners against whom specific crimes can be charged should be brought to justice immediately. Charges of crimes against persons, such as murder, should be made, if possible, rather than charges of

crimes directly affiliated with the resistance movement which may result in martyrdom and serve as a rallying point for increased irregular activity.

(7) Prisoners charged only with being a member of the irregular force will require intensive reeducation and reorientation while confined. In time, consistent with security, those who have demonstrated a willingness to cooperate may be considered for release under parole. Relocation away from previous comrades may be necessary and provision of a means for earning a living must be considered.

(8) Families of prisoners may have no means of support and a program should be initiated to care for them, educate them in the advantages of law and order, and enlist their aid in reshaping the attitudes and beliefs of their confined family members.

U.S. Army Anti-Guerrilla Warfare Manual

Section V. COMBAT OPERATIONS

General

a. Combat operations are employed primarily against the guerrilla elements of an irregular force. However, underground elements some-times attempt to incite large-scale organized riots to seize and hold areas in cities and large towns ; combat operations are usually required to quell such uprisings.

b. Combat tactics used against guerrillas are designed to seize the initiative and to ultimately destroy the guerrilla force. Defensive measures alone result in an ever-increasing commitment and dissipation of forces and give the guerrillas an opportunity to unify, train, and develop communications and logistical support. A defensive attitude also permits the guerrillas to concentrate superior forces, inflict severe casualties, and lower morale. However, the deliberate use of a defensive attitude in a local area as a deceptive measure may prove effective.

c. Constant pressure is maintained against guerrilla elements by vigorous combat patrolling and continuing attack until they are eliminated. This keeps the guerrillas on the move, disrupts their security and organization, separates them from their sources of support, weakens them physically, destroys their morale, and denies them the opportunity to conduct operations. Once contact is made with a guerrilla unit, it is maintained until that guerrilla unit is destroyed.

d. Surprise is sought in all operations, but against well-organized guerrillas it is difficult to achieve. Surprise may be gained by attacking at night, or in bad weather, or in difficult terrain; by employing small units ; by varying operations in important particulars ; and by unorthodox or unusual operations. Counterintelligence measures are exercised throughout planning, preparation, and execution of operations to prevent the guerrillas from learning the nature and scope of plans in advance. Lower echelons, upon receiving orders, are careful not to alter their dispositions and daily habits too suddenly. Tactical cover and deception plans are exposed to guerrilla intelligence to deceive the guerrillas as to the purpose of necessary preparations and movements.

e. The military force attacks targets such as guerrilla groups, camps, lines of communication, and supply sources. Unlike normal combat operations, the capture of ground contributes little to the attainment of the objective since, upon departure of friendly forces, the guerrillas will reform in the same area. Specific objectives are sought that will force the guerrillas to concentrate defensively in unfavorable terrain, and that will facilitate the surrender, capture, or death of the maximum number of guerrillas.

f. Those guerrilla elements willing to fight in open battle are isolated to prevent escape and immediately attacked. Guerrilla elements which avoid open battle are forced into areas which permit containment. Once fixed in place, they are attacked and destroyed. Consideration should be given to the employment of chemical agents to assist in these actions.

g. When the guerrilla area is too large to be cleared simultaneously, it is divided into subareas which are cleared individually in turn, This technique requires the sealing off of the subarea in which the main effort is concentrated to prevent escape of guerrilla groups. Once a subarea is cleared, the main combat force moves to the next subarea and repeats the process. Sufficient forces remain in the cleared area to prevent the development of new guerrilla groups and to prevent the infiltration of guerrillas from un-cleared areas. Pending the concentration of a main effort in a subarea, sufficient forces are employed to gain and maintain contact with guerrilla units to harass them, and to conduct reaction operations.

U.S. Army Anti-Guerrilla Warfare Manual

Reaction Operations

a. Reaction operations are those operations conducted by mobile combat units, operating from static security posts and combat bases, for the purpose of reacting to local guerrilla activities. When a guerrilla unit is located, the reaction force deploys rapidly to engage the guerrilla unit, disrupt its cohesion, and destroy it by capturing or killing its members. If the guerrilla force cannot be contained and destroyed, contact is maintained, reinforcements are dispatched if needed, and the guerrillas are pursued. Flank elements seek to envelop and cut off the retreating guerrillas. The guerrillas should be prevented from reaching populated areas where they can lose their identity among the people, and from disbanding and disappearing by hiding and infiltration. When escape routes have been effectively blocked, the attack is continued to destroy the guerrilla force. The mobility required to envelope and block is provided by ground and air vehicles and by rapid foot movement.

b. Reaction operations are simple, preplanned, and rehearsed because the majority of actions will be required at night. To gain this end the area and possible targets for guerrilla attack must be known in detail. Common targets include desolate stretches and important junctions of roads and railroads, defiles, bridges, homes of important persons, military and police installations, government buildings, public utilities, public gathering places, and commercial establishments. Primary and alternate points are predesignated for the release of re-action forces from centralized control to facilitate movement against multiple targets. Such points are reconnoitered and are photographed for use in planning and in briefing. Within security limitations, actual release points are used during rehearsals to promote complete familiarity with the area.

Harassing Operations

a. Harassing operations prevent guerrillas from resting and re- grouping, inflict casualties, and gain detailed knowledge of the terrain.

They are executed by extended combat patrols and larger combat units.

Specific harassing missions include reconnaissance to locate guerrilla units and camps; raids against guerrilla camps, supply installations, patrols, and outposts; ambushes; marketing targets; assisting major combat forces sent to destroy guerrilla groups ; and mining guerrilla routes of communication.

b. Harassing operations are conducted night and day. Operations at night are directed at guerrillas moving about on tactical and administrative missions. Operations during the day are directed at guerrillas in their encampments while resting, regrouping, or training.

Denial Operations

a. Operations to deny guerrilla elements contact with, and support by an external sponsoring power, are initiated simultaneously with other measures. Denial operations require effective measures to secure extensive border or seacoast areas and to preclude communications and supply operations between a sponsoring power and guerrilla units.

b. The method of contact and delivery of personnel, supplies, and equipment whether by air, water, or land must be determined at the earliest possible time. Border areas are secured by employing border control static security posts, reaction forces, ground and aerial observers, listening posts equipped with electronic devices, and patrols. When time and resources permit, wire and other obstacles, minefields, cleared areas, illumination, and extensive informant nets are established throughout the border area. Radio direction finding and jamming, and Navy or Air Force interdiction or blockade elements may be required.

Elimination Operations

a. Elimination operations are designed to destroy definitely located guerrilla forces. A force much larger than the guerrilla force is usually required. The subarea commander is normally designated as overall commander for the operation. The plan for the operation is carefully prepared, and the troops are thoroughly briefed and rehearsed. Approaches to the guerrilla area are carefully reconnoitered. Deception operations are conducted to prevent premature disclosure of the operation.

b. The encirclement of guerrilla forces is usually the most effective way to fix them in position so as to permit their complete destruction.

(1) If terrain or inadequate forces preclude the effective encirclement of the entire guerrilla held area, then the most important part of the area is encircled. The encirclement is made in depth with adequate reserves and supporting elements to meet possible guerrilla attack in force and to block all avenues of escape.

(2) The planning, preparation, and execution of the operation are aimed at sudden, complete encirclement that will totally surprise the guerrillas. The move into position and the encirclement is normally accomplished at night to permit maximum security and surprise. The encirclement should be completed by daybreak to permit good visibility for the remainder of the operation.

(3) Support and reserve units are committed as required to in- sure sufficient density and depth of troops and to establish and maintain contact between units. Speed is emphasized throughout the early phases of the advance to the line of encirclement. Upon arriving on the line of encirclement, units occupy defensive positions. The most critical period in the operation is the occupation of the line of encirclement, especially if the operation is at night. Large guerrilla formations may be expected to react violently upon discovering that they have been encircled. The guerrillas will probe for gaps and attack weak points to force a gap. Escape routes may be

deliberately established as ambushes.

(4) Units organizing the line of encirclement deploy strong patrols to their front. Air reconnaissance is used to supplement ground reconnaissance. Reserves are committed if guerrilla forces succeed in breaking through or infiltrating the line of encirclement.

(5) Once the encirclement is firmly established, the elimination of the guerrilla force is conducted methodically and thoroughly. A carefully controlled contraction of the perimeter is begun, which may be conducted in any one of three ways:

(a) By a simultaneous, controlled contraction of the encirclement.

(a) By driving a wedge through the guerrilla force to divide the area, followed by the destruction of the guerrillas in each subarea.

(b) By establishing a holding force on one or more sides of the perimeter and tightening the others against them.

(6) During any of the foregoing maneuvers the units that advance from the initial line of encirclement must be impressed with the necessity of thoroughly combing every possible hiding place for guerrilla personnel and equipment. Successive echelons comb all the terrain again. Areas that appear totally inaccessible, such as swamps or marshes, must be thoroughly searched. Guerrilla ruses discovered during the operation are reported promptly to all participating units and agencies. All local individuals, including men, women, and children, found in the area are held in custody and are released only after identification and on orders from appropriate authority.

a. Lack of time, inadequate forces, or the terrain may prevent encirclement operations. Surprise attacks followed by aggressive pursuit may prove successful in these cases. The position, probable escape routes, and strength of the guerrilla forces must be ascertained before launching the operation. Ambushes should be established early on possible escape routes. Patrolling should be conducted in a manner designed to confuse the guerrillas as to specific plans or intentions. Chances of achieving surprise are increased by using airmobile or airborne forces, and by inducing trustworthy local guides who are thoroughly familiar with the terrain and guerrilla disposition to collaborate azid guide the attacking force over concealed routes.

b. After a successful attack on a guerrilla formation, the area is combed for concealed guerrilla personnel and equipment. Documents and records are collected for intelligence analysis. Ambushes are retained along trails in the area for extended periods to kill or capture escapees and stragglers from the guerrilla force.

U.S. Army Anti-Guerrilla Warfare Manual

Combat in Urban Areas

a. Underground elements in cities and towns often incite organized rioting, seize block wide areas, erect street barricades, and resist any attempts to enter the area. Non participants caught in the area are usually held as hostages. The objectives of these operations are to commit the countering force to actions against the civil population which will result in a gain of sympathizers for the irregular force and make it appear that the irregular force is promoting a popular cause.

b. When an urban area has been seized it must be reduced as soon as possible to prevent an apparent success or victory by the irregular force, to maintain popular support for the friendly cause, and to free troops for use elsewhere. The operations required to reduce it resemble normal street and house-to-house fighting. The following tactics are employed:

U.S. Army Anti-Guerrilla Warfare Manual

(1) A cordon is established to surround and seal the barricaded area. The cordon is established at the next street or road, out from the barricaded area, which offers good visibility, fields of fire, and ease of movement. All unauthorized personnel are cleared from the intervening area. The cordon controls all movements into and out of the encircled area.

(2) Announcement is made to the insurgents by such means as loudspeakers and leaflets, that the area will be attacked at a given time unless they lay down their arms, return their hostages safely, and surrender peacefully. Amnesty and protection may be offered to those who surrender prior to the attack.

(3) Maneuver and fire elements attack at the stated time and clear the area as rapidly as possible, with a minimum of killing and destruction of property. The cordon remains in place to maintain security, support the attack by fire where possible, and receive prisoners and rescued hostages from the attacking elements.

(4) If the area is large it is divided into sectors for control purposes. As each sector is cleared, the cordon moves in to exclude it; close surveillance of cleared areas is maintained in case underground passageways are used as escape routes. Succeeding sectors are attacked and cleared one at a time

Cordon and screening posts; one block out.

Crosshatch is barricaded area.

Lettered subsectors are reduced in turn.

High buildings (1 and 2) are seized first.

CHAPTER 3

ORGANIZATION AND EMPLOYMENT OF FORCES

General

a. The operational area, military forces, civil forces, and the population must be organized to provide:

U.S. Army Anti-Guerrilla Warfare Manual

(1) Military or civil area administration.

(2) Static security posts and combat bases.

(3) Security detachments for protecting critical military and civil installations, essential routes of communication, and key communities.

(4) Task forces for conducting police operations against under- ground elements.

(5) Task forces for conducting combat operations against guerrilla elements.

(6) Civil self-defense units for protecting individual villages and small towns.

a. The operational area is subdivided into geographic sectors, or sectors coinciding with internal political subdivisions. Specific sector responsibility for administration and local operations should be arranged for delegation to a single authority, either military or civil.

b. Static security posts are established to protect installations, routes of communication, and communities; maintain control in rural areas and as bases for local reaction operations. No attempt is made to cordon or cover an area with strong points as this immobilizes forces, surrenders the initiative to the irregular force, and invites defeat in detail. Static security posts are organized to be as self-sufficient as possible, reducing dependence on vulnerable land routes of communication.

c. Combat bases are established as needed to facilitate administration and support of company and battalion-size combat units. Combat bases are located within or are immediately adjacent to the units' area of operations and are placed within established static security posts when practicable. A combat base is moved as often as is necessary to remain within effective striking range of guerrilla elements.

Air and ground vehicles are employed extensively for deployment and support of troops to reduce the number of required combat bases.

d. All static security posts and combat bases are organized as both tactical and administrative entities to facilitate local security, rapid assembly, administration, and discipline.

Military Forces

a. The initial force assigned to combat an irregular force must be adequate to complete their elimination. Initial assignment of insufficient forces may ultimately require use of a larger force than would have been required originally. The size and composition of the force will depend on the size of the area, the topography, the civilian attitude, and the hostile irregular force.

U.S. Army Anti-Guerrilla Warfare Manual

Historically, required forces have ranged from a company to a large field army. Organization will usually require the forming of battalion or battle group size task forces which will be given an area responsibility

b. Infantry, armored cavalry, and airborne units are the TOE units best suited for combat against guerrillas. However, many other military units, when reequipped and retrained, can be employed effectively.

c. In active war situations, combat units withdrawn from the line for rest and rehabilitation, or fresh units preparing for commitment should not be assigned a counter-irregular force mission except in emergencies.

d. A mobile force is based at each static security post and combat base which is capable of rapidly engaging reported hostile elements or reinforcing other friendly forces. This force, called a "reaction force," ranges in size from a reinforced platoon to a reinforced company and is capable of rapid movement by foot, truck, or aircraft.

e. The extensive use of patrols is required to assist in local security of installations and to locate and keep the irregular force on the defensive.

(1) Patrols are used in urban and rural areas to augment or re- place civil police and their functions. These patrols may vary from two men to a squad in size. Military police units are ideally suited for employment in such a role and light combat units are quite capable of performing the same functions with little additional training.

(2) Regular combat patrols are formed and employed in a conventional manner and for harassing operations in areas of extensive guerrilla activity.

(3) Extended combat patrols are employed in difficult terrain some distance from static security posts and combat bases.

They are normally delivered into the objective area by aircraft. Extended combat patrols must be capable of employing guerrilla tactics and remaining committed from 1 to 2 weeks, being supplied by air for the period of commitment, and equipped to communicate with base, aircraft, and adjacent patrols. Such patrols may vary from squad to company in size and have the mission of conducting harassing operations and making planned searches of the area so that it leaves no secure areas in which guerrilla forces can rest, reorganize, and train. The effectiveness of an extended combat patrol may be increased immeasurably by appropriate civilian augmentation to include local guides, trackers, and representatives of the civil police or constabulary.

U.S. Army Anti-Guerrilla Warfare Manual

Civil Forces and Local Individuals

a. To minimize the requirement for military units, maximum assistance is sought from, and use is made of civil police, paramilitary units, and local individuals who are sympathetic to the friendly cause. The use and control of such forces is predicated upon national and local policy agreements and suitable screening to satisfy security requirements. Careful evaluation is made of their capabilities and limitations so as to realize their full effectiveness. I. When policy and the situation permit, local individuals of both sexes who have had experience or training as soldiers, police, or guerrillas, should be organized into auxiliary police, and village self-defense units. Those without such experience may be employed individually as laborers, informants, propaganda agents, guards, guides and trackers, interpreters and translators.

b. Civil forces will usually require assistance and support by the military force. Assistance is normally required in an advisory capacity for organization, training, and the planning of operations. Support is normally required in supplying arms, ammunition, food, transportation, and communication equipment.

c. Local and regional police are employed primarily to assist in establishing and maintaining order in urban areas. Local police are most effective in areas which are densely populated.

d. The local defense of communities against guerrilla raids for supplies and terroristic attack may be accomplished in whole, or in part, by organizing, equipping, and training self-defense units. Self-defense units are formed from the local inhabitants and organization is based on villages, counties, and provinces. A self-defense unit must be capable of repelling guerrilla attack or immediate reinforcement must be available to preclude loss of supplies and equipment to guerrillas.

U.S. Army Anti-Guerrilla Warfare Manual

e. Gendarmerie or other national paramilitary units are particularly effective in the establishment and maintenance of order in rural and remote areas. Because of their organization, training, and equipment, they may also be employed in small scale combat operations.

f. Friendly guerrilla units that have operated in the same area as the hostile guerrilla units are usually willing to assist in the counter-guerrilla effort. Such units may be effectively employed in extended combat patrol harassing missions. In addition, members of friendly guerrilla units can serve as trackers, guides, interpreters, translators, and espionage agents, and can man observation posts and warning stations. When friendly guerrilla units are employed, they must be sup-ported logistically and should be subordinate to the military force commander who maintains control and communication by furnishing a liaison party to remain with the friendly guerrilla force, and by controlling the support furnished. Special Forces operational detachments are ideally suited for this purpose.

CHAPTER 4

SPECIAL CONSIDERATIONS

Section I. INTELLIGENCE

General

a. Accurate, detailed, and timely intelligence is mandatory for successful operations against irregular forces. The irregular force is normally ever changing, compartmented, and difficult to identify, and it usually conducts extremely effective intelligence and counterintelligence programs. In consequence, a larger number of intelligence and counterintelligence personnel are often needed than would be required for normal combat operations. The nature of the enemy, the tactical deployment of troops, and the presence of both friendly and hostile civilians in the area dictate modification of normal collection procedures.

U.S. Army Anti-Guerrilla Warfare Manual

b. Intelligence activities are characterized by extensive coordination with, and participation in police, detection, and penetration type operations such as:

(1) Search and seizure operations.

(2) Establishing and operating checkpoints and roadblocks.

(3) Documentation of civilians for identification with central files.

(4) Censorship.

(5) Physical and electronic surveillance of suspects and meeting places.

(6) Maintenance of extensive dossiers.

(7) Use of funds and supplies to obtain information.

U.S. Army Anti-Guerrilla Warfare Manual

(8) Intensive interrogation of captured guerrillas or underground suspects.

a. The conduct of intelligence operations of this nature requires an intimate knowledge of local customs, languages, cultural background, and personalities. Indigenous police, security, military and governmental organizations are usually the best available source of personnel having this knowledge. Individual civilian liaison personnel, interpreters, guides, trackers, and clandestine agents are normally required. Key personnel must be trained in the proper use of indigenous liaison personnel and interpreters. The loyalty and reliability of such persons must be firmly established and periodic checks made to guard against their subsequent defection.

b. Special effort is made to collect information that will lead to the capture of irregular force leaders, since they play a vital part in maintaining irregular force morale and effectiveness.

c. Only when the military forces' knowledge of the terrain begins f o approach that of the irregular force can it meet the guerrillas and the underground on equal terms. Terrain information is continuously collected and processed, and the resulting intelligence is promptly disseminated. Current topographic and photo maps are maintained and reproduced. Terrain models are constructed and used to brief staffs and troops. Military map coverage of the area will frequently be in-adequate for small unit operations. Intelligence planning should provide for suitable substitutes such as large scale photo coverage. Particular effort is made to collect information of:

(1) Areas likely to serve as guerrilla hideouts. Such areas usually have the following characteristics:

(a) Difficulty of access, as in mountains, jungles, or marshes.

(2) Concealment from air reconnaissance.

(a) Covered withdrawal routes.

(b) Located within 1 day's foot movement of small settlements that could provide food, intelligence, information, and warning.

(c) Adequate water supply.

(d) Adjacent to lucrative targets.

(3) Roads and trails approaching, traversing, and connecting suspected or known guerrilla areas.

(4) Roads and trails near friendly installations and lines of communications.

(5) Location of fords, bridges, and ferries across water barriers.

U.S. Army Anti-Guerrilla Warfare Manual

(6) Location of all small settlements and farms in or near suspected guerrilla areas.

(7) When irregular force elements are known or suspected to have contact with an external power: location of areas suitable for airdrop or landing zones, boat or submarine rendezvous, and roads and trails leading into enemy-held or neutral countries friendly to the irregular force.

(8) Dossiers on leaders and other key members of the irregular force should be maintained and carefully studied. Frequently the operations of certain of these individuals develop a pattern which, if recognized, may aid materially in the conduct of operations against them. Efforts are made to obtain rosters and organization data of irregular force elements.' The names and locations of families, relatives, and friends of known members are desired. These i:)ersons are

valuable as sources of information, and traps can be laid for other members contacting them. In communities friendly to guerrillas, some persons are usually engaged in collecting food and providing other aid such as furnishing message drops and safe houses for guerrilla couriers. Every effort is made to discover and apprehend such persons ; however, it may be preferable in certain cases to delay their arrest in order to watch their activities and learn the identity of their contacts. It is sometimes possible to recruit these persons as inform-ants, thereby gaining valuable information concerning the irregular force organization and its communication system.

Overt Collection

a. A large part of the intelligence required for operations against irregular forces is provided by intelligence personnel, troop units, and special information services assisted by civilian agencies and individuals.

U.S. Army Anti-Guerrilla Warfare Manual

b. Reconnaissance and surveillance is an indispensable part of operations against the guerrilla elements. Great care must be exercised, however, so that such activities do not alert the guerrillas and warn them of planned operations. Reconnaissance missions, whenever possible, should be assigned to units with a routine mission which is habitually executed in the area and which they can continue at the same time in order not to arouse the suspicions of the guerrillas of forthcoming operations. Extensive use is made of aerial surveillance using all types of sensors, which means for speedy exploitation of the interpreted results of such coverage.

c. Every soldier is an intelligence agent and a counterintelligence agency when operating against an irregular force. Each man must be observant and alert to everything he sees and hears. He reports anything unusual which concerns the civil population and the irregular force, no matter how trivial.

U.S. Army Anti-Guerrilla Warfare Manual

Covert Collection

Covert collection means are a necessary source of information. Every effort is made to infiltrate the irregular force with friendly agents. Indigenous agents are usually the only individuals capable of infiltrating an irregular force. Such agents are carefully screened to insure that they are not double agents and that they will not relate information gained about friendly forces to the irregular force. The most intensive covert operations possible must be developed consistent with time, available means, and established policy. Agents are recruited among the local residents of the operational area. They have an intimate knowledge of the local populace, conditions, and terrain, and often have prior knowledge of, or connections with members of the irregular force.

Counter- intelligence

a. Irregular forces depend primarily upon secrecy and surprise to compensate for the superior combat power available to the countering military force. Since the degree of surprise achieved will depend largely on the effectiveness of the intelligence gained by the irregular force, intensive effort must be made to expose, thwart, destroy, or neutralize the irregular force intelligence system.

b. Counterintelligence measures may include

(1) Background investigation of personnel in sensitive assignments.

(2) Screening of civilian personnel employed by the military.

(3) Surveillance of known or suspected irregular force agents.

(4) Censorship or suspension of civil communications.

(5) Control of civilian movement as required.

(6) Checks on the internal security of all installations.

(7) Indoctrination of military personnel in all aspects of security.

(8) The apprehension and reemployment of irregular force agents.

(9) Security classification and control of plans, orders, and reports.

a. Counterintelligence operations are complicated by the degree of reliance which must be placed on local organizations and individuals, the difficulty in distinguishing between friendly and hostile members of the population, and political considerations which will frequently hinder proper counterintelligence operation.

Section 11. LOGISTICS

Supply and Maintenance

a. Supply planning lacks valid experience data for the wide variety of environments in which these type operations may occur. Consumption factors, basic loads, stockage levels, and bases of issue must be adjusted to fit the area of operation. Similar factors must be developed for civil forces which may be supported in whole or part from military stocks. The possible need for special items of material must be taken into consideration early.

b. Local procurement should be practiced to reduce transportation requirements.

c. It frequently will be necessary to establish and maintain stockage levels of supply at echelons below those where such stockage is normally maintained. Static security posts and combat bases are examples of localities where stockage will be necessary on a continuing basis.

d. The military force must be prepared to provide essential items of supply to civilian victims of irregular force attack, isolated population centers, and groups which have been relocated or concentrated for security reasons. These supplies may initially be limited to class I items and such other survival necessities as medical supplies, clothing, construction materials, and fuel.

e. Supervision of the distribution of indigenous supplies destined for civilian consumption is necessary.

Assembly, storage, and issue of those items to be used by the civilian population which could be used by irregular forces must be strictly controlled. Local civilians are employed in these functions to the maximum extent possible, but in some situations all or part of the effort may fall on the military agencies. Civil affairs units are organized and trained for this purpose.

f. Security of supply installations is more critical than in normal combat operations. Not only must supplies be conserved for friendly consumption, but their use must be denied to irregular force elements.

Supply personnel must be trained and equipped to protect supplies against irregular force attack, and guard against pilferage of sup- plies by the civilian population.

g. Preventive maintenance should receive emphasis because the very nature of operations precludes the use of elaborate maintenance sup- port. Time is required before and after each mission to effect repair and replacement.

h. Direct support units (DSU), or elements thereof, must provide rapid maintenance support at each static security post and combat base. Although emphasis is upon repair by replacement (direct ex- change) , effort is made to repair items without complete overhaul or rebuild. Stockage of float items is planned to insure that only fast-moving, high-mortality, and combat essential items are stocked.

U.S. Army Anti-Guerrilla Warfare Manual

i. Emergency repair teams, elements of the DSU, are employed to meet special requirements usually experienced in reaction and harassing operations. DSU teams accompany the combat elements and provide on-the-spot minor repair and limited direct exchange.

Transportation

a. Special transportation problems result primarily from

(1) Abnormal distances between static security posts, combat bases, and combat units operating in the field.

(2) Difficult terrain and lack of signal communications in under- developed areas where operations against guerrillas are apt to occur.

(3) The probability that movements of troops and supplies will be subject to attack, harassment, and delay.

Organic transportation means may require augmentation from both military and local sources. Dependent upon

U.S. Army Anti-Guerrilla Warfare Manual

the conditions under which the command is operating, provision of adequate transportation may require such measures as recruiting indigenous bearer units for man pack operations, organizing provisional animal pack units, to include the necessary logistical support, and exploitation of available waterways and indigenous land transportation to include railway and highway equipment.

c. Security will normally be provided all surface movements. Appropriate measures include intensive combat training of drivers and the arming of vehicles involved, aircraft route reconnaissance, and provision of convoy escorts.

d. Aircraft will frequently be the most effective means of resupply because of their speed, relative security from ground attack, and lack of sensitivity to terrain conditions. Army aviation and aviation of other services are utilized. The terrain, tactical situation, and landing area availability may require employment of parachute delivery as well as air-landed delivery.

Evacuation and Hospitalization

a. Medical service organization and procedures will require adaptation to the type operations envisioned. Medical support is complicated by:

(1) The distances between a number of installations where support must be provided.

(2) The use of small mobile units in independent or semi-independent combat operations in areas through which ground evacuation may be impossible or from which aerial evacuation of patients cannot be accomplished.

(3) The vulnerability of ground evacuation routes to guerrilla ambush.

1). Measures that may overcome the complicating factors are

U.S. Army Anti-Guerrilla Warfare Manual

(1) Establishment of aid stations with a treatment and holding capacity at lower echelons than is normal. Such echelons include static security posts and combat bases. Patients to be evacuated by ground transport will be held until movement by a secure means is possible.

(2) Provision of sufficient air or ground transportation to move medical elements rapidly to establish or reinforce existing treatment and holding installations where patients have been unexpectedly numerous.

(3) Maximum utilization of air evacuation. This includes both scheduled and on-call evacuation support of static installations and combat elements in the field.

(4) Provision of small medical elements to support extended combat patrols.

(5) Assignment of specially trained enlisted medical personnel who are capable of operating medical

treatment facilities for short periods of time with a minimum of immediate supervision.

(6) Formation of indigenous litter bearer teams.

(7) Strict supervision of sanitation measures, maintenance of individual medical equipment, and advanced first aid training throughout the command.

(8) Increased emphasis on basic combat training of medical service personnel; arming of medical service personnel ; and use of armored carriers for ground evacuation where feasible.

(9) Utilization of indigenous medical resources and capabilities when available and professionally acceptable.

U.S. Army Anti-Guerrilla Warfare Manual

Construction

a. The underdeveloped transportation system and the difficult terrain conditions normal to areas in which operations against irregular forces may be conducted will frequently require greater light construction than normal combat operations by a similar size command.

Construction planning should provide for:

(1) Combat bases, static security posts and their defenses.

(2) An adequate ground transportation system.

(3) Extensive airstrips, airfields, and helicopter pads to support both Army aviation and aviation of other services.

(4) Essential construction of resettlement areas.

(5) Required support to the local population in civic action projects.

a. The scope of the construction effort requires maximum exploitation of local labor and materiel resources. Combat units also may be required to participate in the construction of facilities both for their own use and for use by the local population.

Section III. SIGNAL COMMUNICATION

General

a. The extreme dispersion of units in operations against irregular forces places a strain on communications means throughout a command. The distances to be covered are greater than the normal area communication responsibility. Augmentation by signal teams and equipment are invariably required to answer basic needs.

b. Radio communication is the primary means, and radio nets are established between all echelons and as needed between military and civil agencies. Ground-air radio communication is established for all airborne, airmobile, and air-supported ground operations.

c. Radio relay and retransmission stations are often required. Air- craft may be used effectively for temporary relay of radio traffic to support a specific short-term action. Ground relay stations must be protected against irregular force raids and sabotage and should be located when possible in areas or installations which are already secured so as to reduce the requirements for guards.

d. Wire communication is used to the maximum extent within secure areas and installations. However, wire communication in unsecured areas is extremely vulnerable to irregular force action and normally will be unreliable. When wire lines extend into unsecured areas and have to be employed, maintenance crews must be accompanied by security guards. A widely used guerrilla tactic is to cut lines and then booby trap the area or ambush the maintenance crew.

e. Visual means of communication can be employed effectively between small units to supplement radio communication and for ground-to-air signaling and marking. The use of flags, lights, smoke, pyrotechnics, heliographs, and panels finds considerable application in such operations.

f. Messenger service between installations or units is limited to air messenger service and motor messengers who travel with security guards or armed convoys. Individual messengers are prime targets for irregular force attack.

g. Police, public, and commercial facilities, and private radio stations and operators are utilized when possible within policy and security requirements.

U.S. Army Anti-Guerrilla Warfare Manual

h. Pigeons may be used to back up electrical means of communication and as a primary means by isolated individuals such as intelligence agents.

Communication Equipment

a. Additional radios are required in most situations to meet basic communication requirements. It is seldom that additional telephone, teletype, or facsimile means will be needed.

b. Additional requirements for equipment are determined based on the distances between units, terrain, and the operations plan as follows:

(1) Short-range, portable FM radios for ground-to-ground and ground-to-Army aircraft communications.

(2) Medium range, portable and mobile AM radios for ground- to-ground communication.

(3) Short-range, portable AM-UHF radios for ground-to-Air Force or Navy aircraft communication.

(4) Appropriate radio relay stations.

(5) Signaling flags and lights.

(6) Panel sets for ground-to-air signaling and marking.

Communication Procedures

a. Irregular forces normally possess neither the sophisticated equipment nor the training required to conduct communication intelligence operations or electronic countermeasures. Nonetheless, normal communication security precautions must be practiced since an irregular force must be credited with the capability of tapping wire circuits, monitoring radio transmissions, and receiving information from a sponsoring power or a conventional enemy force that can conduct communication intelligence operations.

U.S. Army Anti-Guerrilla Warfare Manual

b. All communication facilities are considered important targets by irregular forces and must be protected from sabotage or guerrilla attack, both from within and without.

Electronic Countermeasures

Large, well-developed irregular forces normally depend on radio communication for communicating with a sponsoring power and for control and coordination between major elements. In addition, radios and radar beacons may be employed by an irregular force to communicate with and to guide resupply aircraft, boats, and submarines. Maximum effort is made to: Intercept transmissions for communication intelligence purposes, locate irregular force elements by direction finding, deceive or mislead by false transmissions, locate rendezvous points and drop or landing zones used for resupply missions, or jam their radio transmissions when desirable.

U.S. Army Anti-Guerrilla Warfare Manual

Section IV. SUPPORT BY OTHER SERVICES

Air Force Support

a. Air Force units can assist in the conduct of operations by preventing air delivery of leaders, couriers, equipment, and supplies by a sponsoring power; by aerial resupply and other logistic support functions; by delivering airborne forces to gain tactical surprise; and by conducting close air support, interdiction, air defense, and tactical air reconnaissance, as required.

b. Close air support and interdiction may be difficult to provide and of little value because of the guerrilla capability for dispersion, effective camouflage, moving and fighting at night, and his tactics of clinging to his enemy or of mingling with the populace. Satisfactory results can be achieved, however, when air support can react promptly and attack observed guerilla elements under the guidance of forward air controllers, either on the ground or airborne over the objective area.

U.S. Army Anti-Guerrilla Warfare Manual

c. Well-developed guerrilla forces may have a limited air defense capability which when carefully concealed and moved often, can re- duce the effectiveness of air support. Another possible capability of guerrillas is the use of deceptive radio navigation transmitters or other deception measures to misdirect aircraft. Because of such possibilities, intelligence reports are carefully screened for indications of changes in guerrilla capabilities.

Navy and Marine Support

a. Navy Forces. Navy forces can assist in operations against irregular forces by disrupting such irregular force supply channels as are maintained by coastal craft; by providing sea transport for rapid concentration of ground forces when opportunities are presented to attack guerrilla formations in areas contiguous to the sea; by shore bombardment to assist ground operations in areas adjacent to the sea and by preventing the seaward escape of irregular force elements.

Navy forces can also provide seaborne resupply and other logistic support functions.

b. Marine Forces. Marine forces can assist in operations against irregular forces by conducting operations both on the ground and in the air in a manner similar to both Army and Air Force forces.

CHAPTER 5

TRAINING - General

a. All troops committed to operations against irregular forces must be trained to appreciate the effectiveness of irregular forces and the active and passive measures to be employed. Troops must be indoctrinated never to underrate guerrillas. To look down on guerrilla forces as inferior, poorly equipped opponents breeds carelessness which can result in severe losses.

6. Training for operations against irregular forces is integrated into field exercises and maneuvers as well as

U.S. Army Anti-Guerrilla Warfare Manual

in individual and small unit training programs. Aggressor force tactics in training exercises should include irregular activities, both covert and overt. Normal individual and small unit training should emphasize:

(1) Physical conditioning

(2) Tactics and techniques appropriate to urban areas, mountains, deserts, swamps, and jungles.

(3) Tactics and techniques of CBR warfare.

(4) Extended combat patrol operations utilizing only such supplies as can be transported by the patrol.

(5) Immediate reaction to unexpected combat situations.

(6) Employment of Army aviation, to include techniques of airmobile assault and casualty loading.

(7) Aerial resupply by Army and Air Force aircraft to include drop and landing zone marking and materiel recovery techniques.

(8) Night operations.

(9) Techniques of raids, ambushes, ruses, and defensive and security measures against these types of operations.

(10) Riot control to include employment of chemical agents.

(11) Police-type patrolling and the operation of roadblocks and checkpoints.

(12) Cross-training on all communications equipment available within the type unit and in communication techniques.

(13) Cross-training on all individual and crew served light weapons available within the type unit.

(14) Marksmanship.

(15) Target identification.

c. When the characteristics of the area of operations and the irregular forces are known, further specialized training is required in such specially applicable subjects as:

(a) The use of animal transport for weapons and logistical support.

(b) Movement techniques, field craft, and improvisation for fighting and living in mountains, jungles, or swamps.

(c) Cold-weather movement including ski and sled operations.

(c) Utilization of water means to gain access into areas occupied by irregular forces.

(d) Survival techniques to include manner and technique of living off the land for short periods.

(e) Indoctrination in the ideological and political fallacies of the resistance movement.

(f) Cross-country movement at night and under adverse weather conditions to include tracking and land navigation.

(g) Police-type search-and-seizure techniques, counterintelligence, and interrogation measures.

(h) Convoy escort and security.

(i) Advanced first aid, personal hygiene, and field sanitation.

(j) Use and detection of mines, demolitions, and booby traps.

() Small-unit SOP immediate-action drills.

a. Prior to entry into an area of operations, all troops must receive an orientation on the nature of the terrain and climate, unusual health hazards, customs of the population, and their relations with the civil population.

b. Training for specific operations often requires detailed rehearsal.

Rehearsals are conducted in a manner which will not compromise actual operations, but are conducted under like conditions of terrain and time of day or night.

Morale and Psychological Factors

a. Troops employed against irregular forces are subjected to morale and psychological pressures different from those normally present in regular combat operations. This is particularly true in cold war situations and results to a large degree from:

b. The ingrained reluctance of the soldier to take repressive measures against women, children, and old men who usually are active in both overt and covert irregular activities or who must be resettled or concentrated for security reasons.

(2) The sympathy of some soldiers with certain stated objectives of the resistance movement such as relief from oppression.

(3) Fear of the irregular force due to reported or observed irregular force atrocities and conversely, the impulse to take vindictive retaliatory measures because of such atrocities.

(4) The characteristics of the operations to include:

U.S. Army Anti-Guerrilla Warfare Manual

(a) The difficulty in realizing or observing tangible results in arduous and often unexciting operations.

(b) The primitive living and operating conditions in difficult terrain.

(c) The long periods of inactivity which may occur when troops are assigned to static security duty.

(5) Inexperience in guerrilla and subversive tactics.

(6) Ingrained dislike of clandestine and police-type work.

a. Soldiers who are untrained in such operations are prone to bewilderment when faced by irregular force tactics and the intense political and ideological feelings of guerrillas.

b. Commanders at all echelons must carry out, on a continuing basis, an indoctrination, education, and training program which will effectively offset these

morale and psychological pressures. In addition,

the training program must insure that troops impress the

local populace with their soldierly ability, courtesy, and

the neatness, efficiency, and security of their person,

camps and installations.

MIKAZUKI PUBLISHING HOUSE
CATALOGUE
Mikazuki Jujitsu Manual
25 Principles of Martial Arts
Karate 360
Political Advertising Manual
Learning Magic
Stories of a Street Performer (Pop Haydn)
Magic as Science & Religion
The Bribe Vibe
World War Water
Small Arms & Deep Pockets
Arctic Black Gold
Find the Ideal Husband
John Locke's 2nd Treatise on Civil Government
The History of Acid Tripping
I Dream In Haiku
Mikazuki Political Science Manual
Tokiwa; A Japanese Love Story
The Card Party; Theater Play
Hagakure; The Book of Hidden Leaves
MMA Coloring Book
DIY Comic Book
Freakshow Los Angeles
Swords & Sails: The Legacy of the Red Lion
Coming to America Handbook
The Medium Writer
California's Next Century 2.0: Economic
Renaissance
Self-Examination Diary: Good/Bad Deeds Log
Master Password Organizer Handbook
George Washington's Farewell Address
Customer Profile Organizer

United Nations Charter
DIY Comic Book Part II
Storyboard Book: Make Your Movie Series
Basketball Team Play Design Book
Football Play Design Book
T-Shirt Design Book
Rappers Rhyme Book: Lyricists Notebook
Japan History Coloring Book
Magicians Coloring Book
The Adventures of Sherlock Holmes
Words of King Darius
The Art of War
The Book of Five Rings
Tao Te Ching
Captain Bligh's Voyage
Beginner's Magician Manual
The Man that Made the English Language
The Arrival of Palloncino
The Irish Republican Army Manual of Guerrilla Warfare
Living the Pirate Code
Van Carlton Detective Agency; The Burgundy Diamond
Quotes Gone Wild
Shogun X the Last Immortal
The Art of Western Boxing
William Shakespeare's Sonnets
U.S. Military Boxing Manual
Mythology Coloring Book
U.S. Army Anti-Guerrilla Warfare Manual

Education Is the Key to Happiness

NOTES

NOTES

NOTES

NOTES

NOTES

NOTES

www.ingramcontent.com/pod-product-compliance
Lightning Source LLC
Chambersburg PA
CBHW060048210326
41520CB00009B/1309